The Future of Quality Jobs

Quality 4.0 New Opportunities for Quality Experts

Published by KIT IN MOTION DESIGN Inc.
Vancouver, Canada

National Library of Canada Publication Data
Shakeri, Sean
The Future of Quality Jobs / Sean Shakeri
– 1st Edition.

ISBN: 978-1-7771357-4-4

1. Quality jobs   2. Future jobs   3. Technology
4. Career development

KITINMOTION.COM

Preface

47% of current jobs will disappear in the next 20 years, say Oxford university researches.

Artificial Intelligence, smart robots, new technologies like machine learning, data science, 5G, 3D printing are going to become more universal in every business.

Quality jobs, including quality control, quality assurance, quality engineer, quality auditor ... will be affected by these new technologies.

New technologies in industrial revolution 4.0 able machines communicate, adjust, calibrate and fix each other!

Although many jobs will disappear or shrink, there are a lot of unique opportunities arising in the market in the next 10 to 20 years and these opportunities might not happen again anytime soon.

The purpose of this book is to advise quality professionals about the skills that they need to have in industrial revolution 4.0 to get high-income jobs.

If you are a quality inspector, engineer, consultant, manager, leader, auditor ... or your job is related to quality, you should read this book.

"The Future of Quality Jobs" is brief, informative, refreshing and straightforward.

Sean Shakeri

Table Of Contents

Introduction

The First
Industrial Revolution

The First Industrial Revolution 1765 – 1890 was characterized by a transition from hand production methods to machines through the use of steam power and waterpower.

The first Industrial Revolution created a major improvement in human life history; almost every aspect of our daily life was influenced in some way.

Especially average income began to increase during this time.

Standard of living for the general population in the western world began to increase during this time.

The first Industrial Revolution also led to an unmatched rise in the rate of population growth.

The Second Industrial Revolution

The second Industrial Revolution 1890 – 1940 was a period of mass production and great economic growth with an increase in productivity.

Many management methodologies were developed during this period of time.

The second Industrial Revolution caused a surge in unemployment since machines replaced many factory workers.

The Third Industrial Revolution

The third Industrial Revolution, also known as the Digital Revolution, occurred in the late 20th century.

Correct, exact, up to date and on time information added value to the processes and products during this revolution.

Amazon, Microsoft, Apple, Facebook are good examples of successful companies during this period of time.

The Fourth Industrial Revolution

The phrase **Fourth Industrial Revolution** was first introduced by Klaus Schwab, executive chairman of the World Economic Forum in 2015.

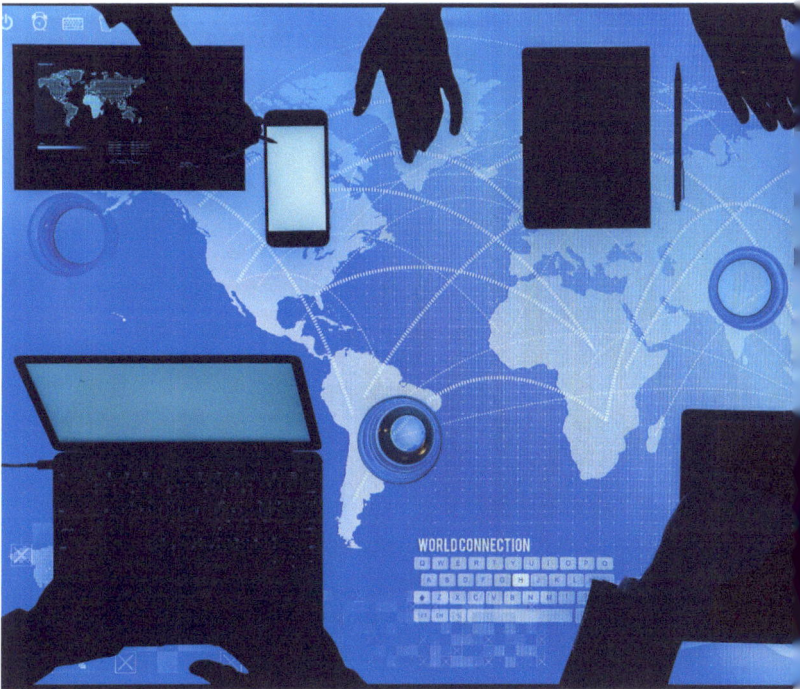

The key feature of this revolution is cyber-physical systems. Businesses have also taken automation to a whole new level.

The prior revolutions mainly focused on economic changes, whereas Industrial

Revolution 4.0 has been causing political and cultural changes in North America and the rest of the world.

Industrial revolution 4.0 is about connectivity and emerging technologies such as robotics, artificial intelligence, nanotechnology, biotechnology, Internet of Things, 5G, and 3D printing.

These new technologies provide a new level of interconnectivity that has never been seen before.

Maybe the biggest difference between industrial revolution 4.0 and other industrial revolutions is "the speed of change."

Meaning top trending inventions like Artificial Intelligence (AI), Virtual Reality, Augmented Reality, and Internet of Things (IoT) are growing to change the world much faster than any other tools and inventions in the past.

Industrial Revolution And Quality Revolution

Quality functions are the part of operational functions. When operation changes, quality functions change simultaneously.

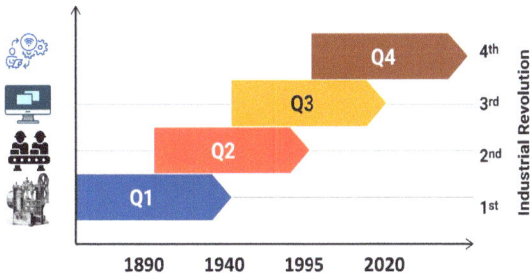

The definition of quality has been changed during Industrial Revolutions.

For example, in the third Industrial Revolution, quality means customer satisfaction whereas in the second industrial revolution, quality was about product measurement and productivity.

Definition Of Quality

Here are the applications of quality in each Industrial Revolution:

Quality 1

- Production volume was more important than quality
- Quality meant simple inspection and separated good products from bad products
- Inspection did not focus on cost reduction

Quality 2

- Primary focus on maximizing productivity
- Quality meant inspection, measurement and using simple analysis to improve processes
- Labour performance started to be measured and considered as productivity

Quality 3

- Quality meant customer satisfaction
- Continual improvement was applied to business processes
- Standardization started for business improvement like ISO9001, ISO14001, ISO18001

Quality 4

- Process and product designers are responsible to design quality in processes
- Machines learn how to self regulate and manage their own productivity and quality
- Human performance in quality is still important but there is a shift from statistical quality tools to business design

Why Most Quality Jobs Are Dying?

What Is Happening In Quality Job Market?

The ultimate goal of any business is creating revenue and optimizing profit, and quality systems are supposed to facilitate this.

It is obvious that business owners are looking for the methods or technologies to decrease operational costs in order to increase profit.

One of the most effective methodologies is finding and eliminating non-added value activities.

A great number of business owners believe that many quality functions do not add any values to businesses in terms of sales and profit. In fact, it absorbs the company's profits!

Functions like inspection, manual data gathering, data entry, corrective actions, and auditing are expensive operations for any businesses and they don't add any values to products or services.

Customers don't care if the companies hire 100 people or 1 person in quality departments to make sure the quality is good.

Customers care if the product is reliable and meets their expectation as the company promised.

As Dr. Joseph Juran said, Quality means "fitness for use".

Up to now, organizations have hired quality experts to do inspections and used statistical methodologies like 6 Sigma because there are no better ways to make sure that products are flawless.

Entrepreneurs believe that quality functions need to be managed and eliminated as much as possible by designing quality in processes at the first place.

Customers or end users might think that having a lot of inspections means better quality of products, but **quality experts know having a lot of inspections means "lack of quality in processes".**

Running a lot of inspections shows that the operation is not reliable and is not under control.

Inspection does not add any values to products. Instead, it increases the time and cost of the operations.

One of the goals for every organization is using affordable technologies to decrease the number of manual inspections.

Many business owners believe that quality experts are not sensitive enough about time and moneymaking activities.

Traditional education system, lack of direct connection between quality management systems and sales numbers, and advanced new technologies like artificial intelligence, robots, automation, and intelligent IT systems, put most quality jobs in danger.

A recent study done by Boston Consulting Group and ASQ shows that quality experts should take actions and equip themselves with new tools and knowledge if they want to stay in quality field in the future.

Here is the comparison between basic and modern quality tools.

Here is the comparison between basic and modern quality tools.

7 Basic Quality Tools

1. Cause-and-effect diagram
2. Check sheet
3. Control chart
4. Histogram
5. Pareto chart
6. Scatter diagram
7. Flow chart or run chart

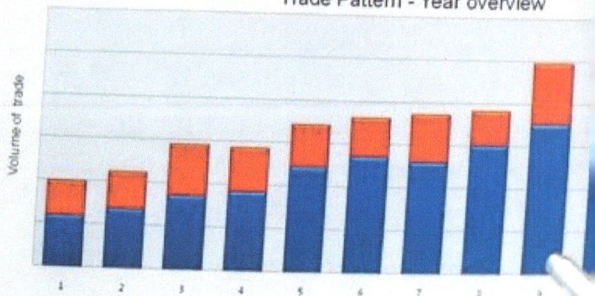

Trade Pattern - Year overview

Volume of trade

1   2   3   4   5   6   7   8   9

Stock Market share

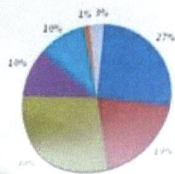

1% 3%
10%
27%
10%

■ 1
■ 2
■ 3
■ 4
■ 5
■ 6
■ 7

Share valu

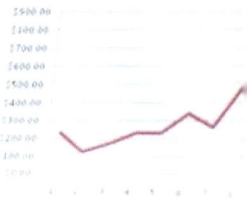

$900.00
$800.00
$700.00
$600.00
$500.00
$400.00
$300.00
$200.00
$100.00
$0.00

7 Future Quality Tools

1. Artificial Intelligence
2. Big data
3. Blockchain
4. Deep learning
5. Enabling technologies
6. Machine learning
7. Data science

A comparison between traditional and modern quality tools shows that quality experts need to have a background in computer science, electronics, IT, communication technology, and psychology.

This is a new challenge for many quality professionals.

Quality 4 is a new generation of quality that might change everything we know about quality.

Smart applications and AI will replace inspection, quality control, and quality management positions.

Inspection, data gathering, analyzing data, finding the best solution, and even making decisions for improving processes will be planned and accomplished by artificial intelligence.

Here is what we could expect from Quality 4:

1. Quality shifts from controlling the processes by operators to the process designers

2. Machines learn how to self regulate and manage their own productivity and quality

The question is, how long does it take to see this in real life?

How much time do we have to adapt to these new technologies?

New Paradigm

In 2018, Boston Consulting Group (BCG) invited ASQ to partner in a study exploring Quality 4.0, focused primarily on the manufacturing sector in the U.S. and Germany.

The study looked at professionals' perceptions of skills needed to implement Quality 4.0, obstacles to implementing, and the usage of digital quality management.

The study shows that 16% of organizations surveyed have implemented Quality 4.0 initiatives.

Also, the report revealed that automation would be at least 10 times faster than previous quality generations.

This is a new paradigm shift and we know there are always opportunities when a paradigm shift in technology happens.

New Opportunities For Quality Experts

Quality 4.0 – New Opportunities For Quality Experts

Skills and years of experience that quality experts acquired during years might not be enough for Quality 4.0.

In the Industrial Revolution 4.0, quality experts need to learn about new technologies and their applications.

New quality tools are not like other tools that have been used by quality experts so far. These new technologies will help quality experts to implement quality systems.

"Quality 4.0 is NOT about new tools; it is about implementation."

New technologies like artificial intelligence, augmented reality, affordable sensors and actuators could help quality experts implement their quality improvement plans.

Industry 4.0 new technologies could be considered as the biggest opportunities for quality professionals who want to learn and grow.

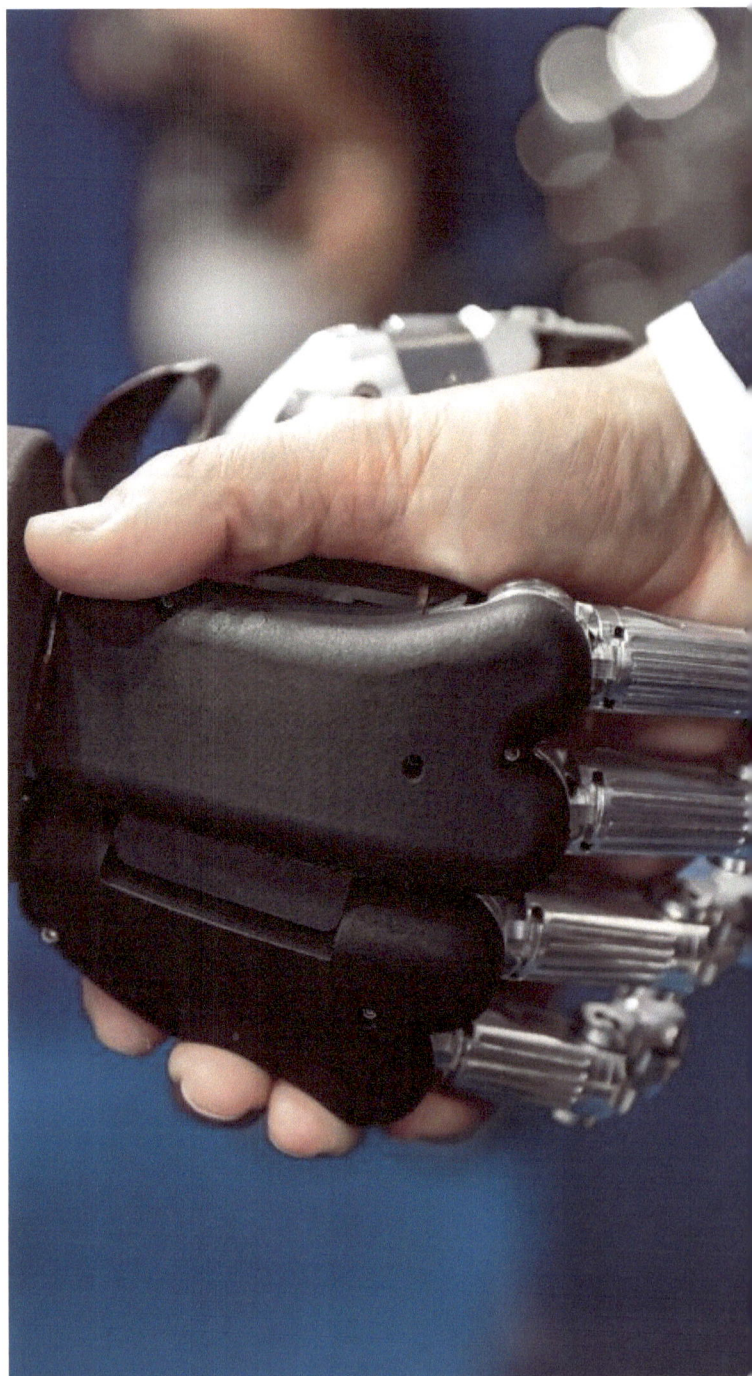

Are You In Or Out?

Using new technologies is not a choice. It is essential for both small and large companies to adapt themselves to industry 4.0 technologies in order to compete in the market.

New technologies are much more affordable compared to 10 years ago.

A company that needed to spend **millions of dollars** for
digitization in 2010, now they **only need thousands of dollars** to do so in 2020.

What actions quality experts with corporate jobs should take?

Where is the opportunity?

How to learn new high-income skills?

Actions Quality Experts Should Take

1. Equipping themselves with Q4 tools

According to the recent research by Boston Consulting Group and ASQ (2019), the number one challenge of implementing Quality 4.0 is the digital skills talent gap.

For frontrunners in Quality 4.0 implementation, data-related barriers are also big challenges for quality experts.

Learning new quality tools is something that most quality experts are familiar with. Most of them love to upgrade their knowledge and learn new skills.

If you love technologies and you are good at computer science or you have a background in IT, this is the option you might want to consider.

Learning about these new technologies takes time and sometimes it is too challenging for experienced quality professionals.

2. Moving to high-level quality positions

New technologies are going to take care of operation in certain industries.

When you are mostly involved in operational tasks like quality control, testing, inspection and measurement, your job is more likely in danger.

Moving from inspection and testing positions to high-level quality positions like designing quality systems, or processes is another solution for quality experts.

To move to a higher position, quality experts need to equip themselves with quality and business system design skills[1].

Your job is safe as long as you are creative, able to design processes or solve complicated problems.

---

[1] kitinmotion.com/sa-program

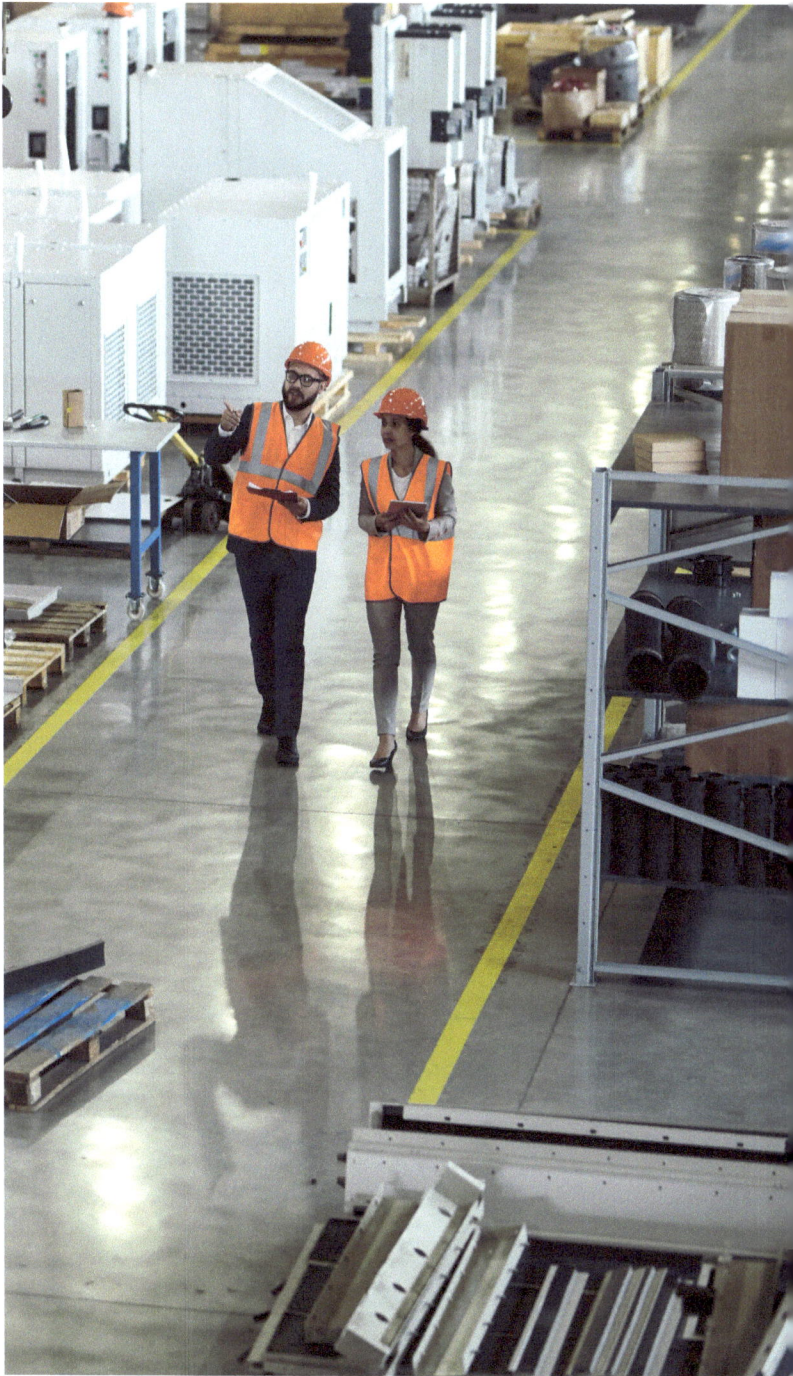

3. Moving to quality positions in other industries

New technologies will not affect all the industries at the same time.

Manufacturing industries will be the center of this change.

Health and food industries have more time to adapt themselves to new technologies because of mandatory safety and health regulations.

Applying for a quality job in a different industry is somehow challenging as most companies ask for relevant experience in their industry.

One of the possible solutions is applying for an entry-level job and then moving to a higher-level position. This might not be an option for experienced quality professionals.

Another possible option is getting knowledge or certificates related to the industry that you are interested in.

Sometimes quality experts can use their network to switch to a new industry.

4. Moving to other positions similar to quality

Thinking about having a new position and moving to a different area of work is not an easy decision.

However, for those experienced quality experts who are not afraid of change and like to challenge themselves, moving to other positions is always an option.

One of the popular options that many experienced quality experts could consider is moving to coaching and consulting businesses.

In that case, quality professionals should equip themselves with interpersonal industry 4.0 soft skills like growth mindset, strategic goal setting, fast problem solving skills, marketing techniques, coaching and consulting methodologies[2].

Also, they might need to learn how to design and develop a business system from scratch, which is not a big deal for most experienced quality experts.

---

[2] kitinmotion.com/sa-program

Which Quality Careers Will Be In Danger?

Quality Experts Know About The Future Changes

Over 450 quality experts joined my webinar, "Why Quality Jobs Are Dying?"

92% of the attendees claimed that they came to the event because of the topic.

Most quality professionals have already known about future challenges and they are ready to take actions to upgrade their career.

Although few experts are still skeptical about the changes that are going to happen in the next 10 to 15 years, studies done by Boston Consulting Group, ASQ, Mckinsey Global Institute, World Economic Forum shows that changes in technology will dramatically affect the way people work.

New technologies allow machines to communicate with each other.

Communication has been the main job for human in Industrial Revolution 3.0 but machines are going to take over part of it in Industrial Revolution 4.0.

Machines can also collect data from operations, analyze it and give a solution to the problems at real time.

This is the job for quality engineers at the moment, although machines could be hundreds of times more efficient and accurate.

Robots can also learn from the processes and improve operational functions.

Mistake Proofing or Poka-Yoke, and Error Proofing have been the main hot topics for years and can be easily accomplished by industry 4.0 new technologies.

Dr. Erik Brynjolfsson, Director of the Massachusetts Institute of Technology believes that machine learning is enabling us to improve machine functions by 1,000,000 times!

"This enables machines to recognize objects, which is something that just humans could do until now."

"This is the moment of choice and opportunity.

This could be the best 10 years ahead of us or one of the worst because we have more power than we had before."
– Dr. Erik Brynjolfsson

Which Quality Jobs Are More In Danger?

In order to answer this question, we need to understand the new skills and qualifications that are highly demanded in Industrial Revolution 4.0.

According to the World Economic Forum, here is the comparison list of the skills needed in 2015 and 2020:

2015

1. **Complex problem solving**
2. Coordinating with others
3. People management
4. Critical thinking
5. Negotiation
6. **Quality control**
7. Service orientation
8. Judgment and decision making
9. Active listening
10. **Creativity**

2020

1. **Complex problem solving**
2. Critical thinking
3. **Creativity**
4. People management
5. Coordinating with others
6. Emotional intelligence
7. Judgment and decision making
8. Service orientation
9. Negotiation
10. **Cognitive flexibility**

What Is The Answer?

The report shows that "Quality Control" is not on the list in 2020 anymore.

"Cognitive flexibility" is the skill that has been moved to the top 10 skills.

Cognitive flexibility has been described as the mental ability to switch between thinking about two different concepts and thinking about multiple concepts simultaneously.

"Creativity" moved from number 10 to number 3, which means your job will be safer if you are doing some kind of creative jobs, like designing processes, procedures or products.

Moreover, "Complex problem solving" is still on the top of the list.

Many companies in industry 4.0 need to have experts to solve the problems that could not be solved by machines.

You can easily make a list of the skills that you are using on a daily basis, and then compare your skills with the ones that are highly demanded in the future job market.

Do you think that your job is safe according to the list that you have compiled?

What Skills Should I learn To Get A High-income Job?

What Skills Should I Learn To Get A High-income Job?

You might think this is not true! Nothing is really happening to my job. My job is safe.

But what is happening in the market right now says the opposite.

Many people lost their jobs and they believe this is the result of the global pandemic.

They might be right but there is another reason that most people don't know, or they don't really want to know!

It is "the **4th Industrial Revolution** and new affordable technologies for almost every industry."

The reality is that COVID-19 has been speeding up the process for this transformation from industry 3.0 to industry 4.0.

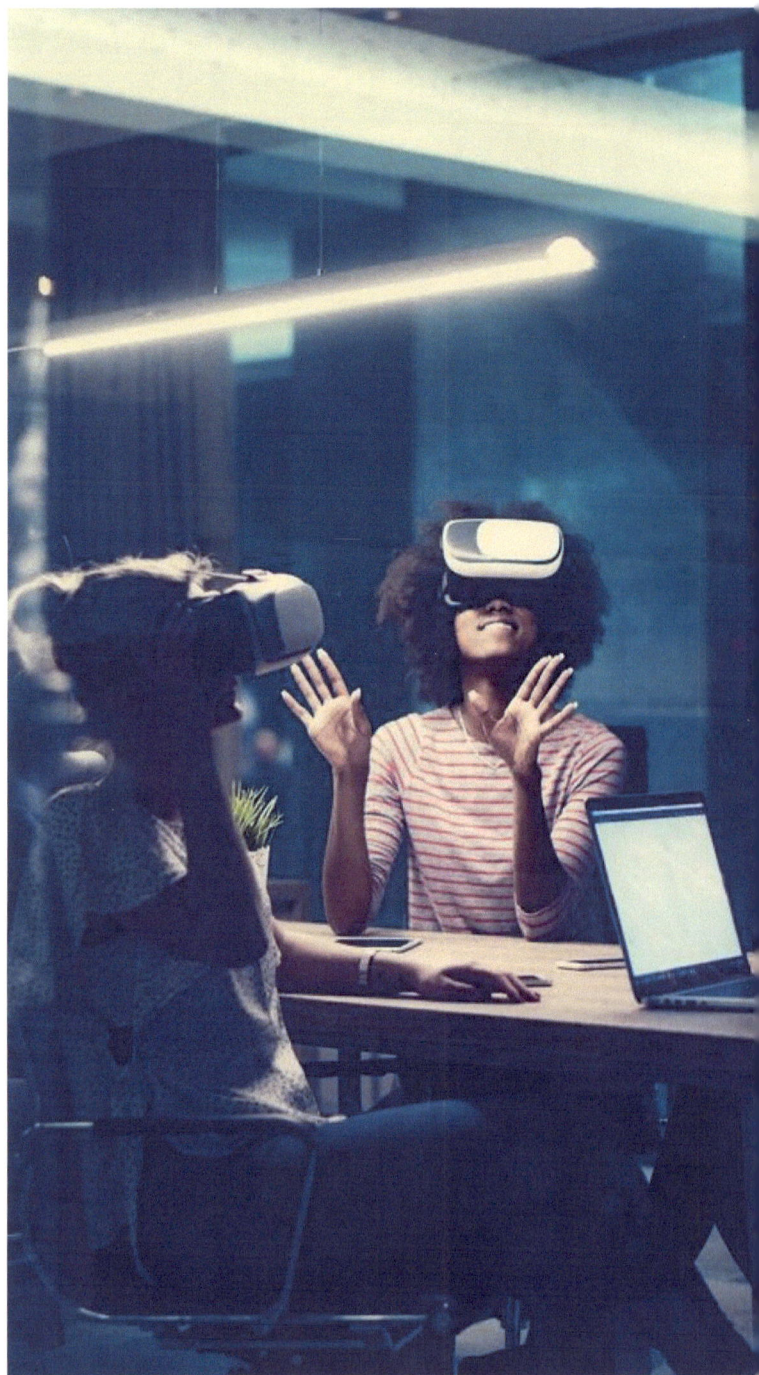

Improve Your Skills To Get High-income Jobs

Industry 4.0 is one of the biggest opportunities in human history for those who want to grow, to be financially stable, and to go to the next step of their career.

There are a lot of new opportunities arising in the market and quality experts can definitely take the advantage.

Let's have a look at the skills that are required for quality 4.0.

Technological Skills

The study done by Boston Consulting Group and ASQ shows that quality experts need to equip themselves with new technological tools to improve their careers and find high-income quality jobs in the future.

Here are the technologies that quality experts need to know in industrial revolution 4.0:

1. Artificial intelligence

Computer vision, language processing, chatbots, personal assistants, navigation, robotics, making complex decisions

2. Big data

Infrastructure (such as MapReduce, Hadoop, Hive, and NoSQL databases), easier access to data sources, tools for managing and analyzing large data sets without having to use supercomputers

3. Blockchain

Increasing transparency and auditability of transactions, monitoring conditions so transactions don't occur unless quality objectives are met

4. Deep learning

Image classification, complex pattern recognition, time series forecasting, text generation, creating sound and art, creating fictitious video from real video, adjusting images based on heuristics

5. Enabling technologies

Affordable sensors and actuators, cloud computing, open-source software, augmented reality (AR), mixed reality, virtual reality (VR), data streaming (such as Kafka and Storm), 5G networks, IPv6, IoT

6. Machine learning

Text analysis, recommendation systems, email spam filters, fraud detection, classifying objects into groups, forecasting

7. Data science

The practice of bringing together heterogeneous data sets for making predictions, performing classifications, finding patterns in large data sets, reducing large sets of observations to most significant predictors, applying sound traditional techniques to generate viable models and solutions

It is understandable that learning these new technologies looks challenging for many quality experts.

However, learning these technologies is much easier than it seems.

Not many quality experts know much about these technologies at the moment, and if you know, you have the advantage of getting new high-income jobs.

Young quality experts have already started to learn these new technologies.

Industry 4.0 Soft Skills

There is no time in modern human history when learning soft skills is more important than professional skills!

New learners are joining online platforms every day to learn more about soft skills.

Here are the most wanted soft skills according to the World Economic Forum:

1. Complex problem solving
2. Critical thinking
3. Creativity
4. People management
5. Coordinating with others
6. Emotional intelligence
7. Judgment and decision making
8. Service orientation
9. Negotiation
10. Cognitive flexibility

Doesn't really matter what job you are going to apply in the future, competition is high when it comes to soft skills.

What Are The High-income Jobs in Industrial Revolution 4.0?

Industrial Revolution 4.0 Universities

There is no doubt that the traditional education system doesn't support students to find high-paying jobs in the future.

For baby boomers (born in 1946 to 1964), having a degree from a good university meant getting a good job.

Generation X (born in 1965 to 1980) needed to have a degree plus professional certificates to find a job.

This has been changed dramatically for millennials (born in 1981 to 1996) and generation Z (born in 1997 to 2012).

Technologies are growing much faster than our education system and by the time students graduate from colleges or universities, their information is outdated and obsolete.

Companies in industry 4.0 won't rely on college degrees or certificates as much as they did in the past.

Many tech companies have already improved their own hiring system to recruit employees based on their skillset certificates.

For example, tech companies like Microsoft have been in the education industry for years and they have been issuing their own certificates.

Google recently announced that they are launching a selection of professional courses, a 6-month program that teaches candidates how to perform in-demand jobs.

These courses, **Google Career Certificates**, help job seekers find high paying jobs almost immediately by teaching them fundamental skills.

Fast-growing tech companies like Google, Microsoft, Amazon, Apple claim their courses, which would cost a fraction of traditional university education, prepare students to immediately find work in high-paying career fields in industry 4.0.

According to Google, here are the average salaries Google offers after completing their courses (no degree or prior experience required):

• **Project manager ($93,000)**

• **Data analyst ($66,000)**

• **UX designer ($75,000)**

What do you think if you can start working for a multibillion-dollar company after finishing a 6-month program?

Industry 4.0 Skillset Certificates

There are a variety of complementary certificates you might need to have for getting a job in Industrial Revolution 4.0 but remember, organizations don't evaluate their candidates based on their certificates as they did in Industrial Revolution 3.0.

We have learned a lot during this global pandemic. Companies look at employees' performances, not their degrees or certificates when they have to go through a layoff!

Having degrees and professional certificates is still a good idea but not enough to find a high-paying job in the future.

Industry 4.0 companies don't pay for degrees or certificates. They pay for employees' performances.

Although Industry 4.0 is about accomplishment not certificates, you might want to improve your skills by enrolling in programs with certificates.

There are three groups of certifications for industry 4.0 in the market:

1. Industry 4.0 fundamental skills
2. Industry 4.0 executive soft skills
3. Technological skills

1. Industry 4.0 fundamental skills

You can apply for programs or courses in essential industry 4.0 skills:

- Growth mindset
- Communication
- Life and career planning
- Coaching skills
- Consulting techniques
- Business design
- Rapid problem solving skills
- Digital marketing
- Copyrighting
- Content creating

These certificates might be a little bit odd for quality engineers who love to work on numbers and statistics but these skills will be part of the job requirements in the future!

2. Industry 4.0 executive soft skills

You can enrol in executive soft skill programs like:

- Complex problem solving
- Critical thinking
- Creativity
- People management
- Coordinating with others
- Emotional intelligence
- Judgment and decision making
- Service orientation
- Negotiation
- Cognitive flexibility

Universities have started having part time or online programs to teach these executive soft skills.

3. Technological skills

Basic technological skills will be essential for future quality professionals. You can register in technological programs like:

- Artificial Intelligence
- Big data
- Blockchain
- Deep learning
- Enabling technologies
- Machine learning
- Data Science

There are various online courses for learning these industry 4.0 technical skills.

Future quality leaders are those who have knowledge and experience in all these three areas.

What About Other Quality Certificates

Although there is still a market for having certificates for quality experts in measuring tools, statistical methods, 6 Sigma, quality auditing, etc., these certificates belong to industrial revolution 3.0 and the demand is decreasing.

Quality management systems are going to merge with business systems and new technologies are going to take care of many quality functions.

ROBO – The Future Quality And Business Manager

What Is ROBO?

ROBO is an imaginary "Quality 4.0 Manager" that is responsible for gathering and analyzing data to improve business operation.

ROBO is a highly smart robot that can communicate with other robots and make decisions for improvement.

ROBO can also monitor and improve hundreds of operations at real time and balance operation system.

ROBO is an expert in root cause analysis and can offer possible solutions in a fraction of second.

ROBO is a great communicator and it connects with other robots and human in different countries through 5G network.

Where Is ROBO Made?

ROBO is made in Japan or China and can be manufactured in Europe and North America as well.

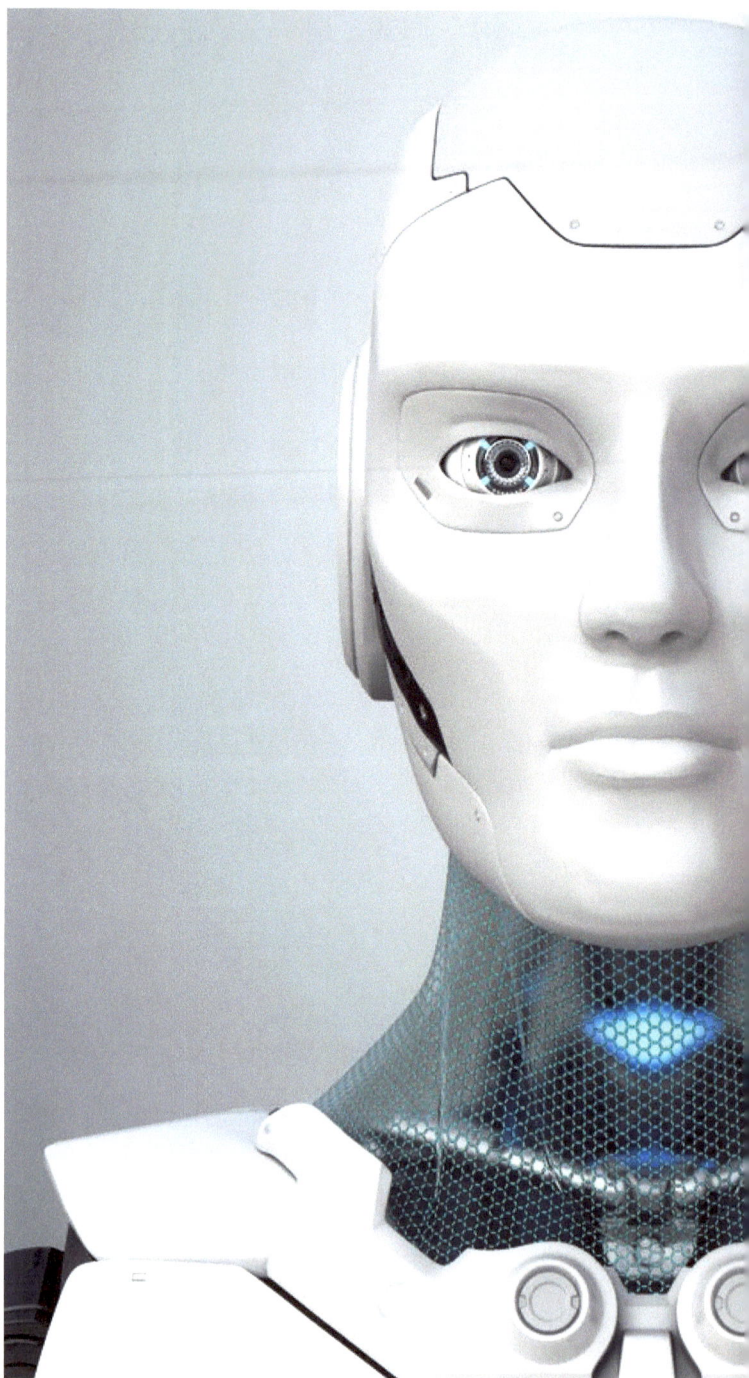

ROBO's Job Description

Qualifications

- Measuring parts, products, processes and analyzing data

- Managing a large team of robots (e.g. 10 to 100 team members)

- Receiving data from other robots, analyzing and processing data, and making adjustments with robots

- Calibrating other robots automatically

- 6 Sigma professional

- Ability to collect and analyze quantitative and qualitative data to generate fact-based recommendations and measurement of progress

- ERP system professional

- Ability to analyze information and determine potential options for solutions to business problems

Interpersonal Skills

- Communication skills – Communicate with other machines wirelessly; communicate with human via voice, messages, emails, videos

- Highly proficient in PowerPoint, Excel, Word

Is ROBO Affordable?

ROBO is a part of the larger system called ROBO system.

Companies need to invest on ROBO system in Industrial Revolution 4.0 when they want to go to the next level.

The new technologies are getting more affordable for businesses as even small businesses in Japan and China have started using these systems.

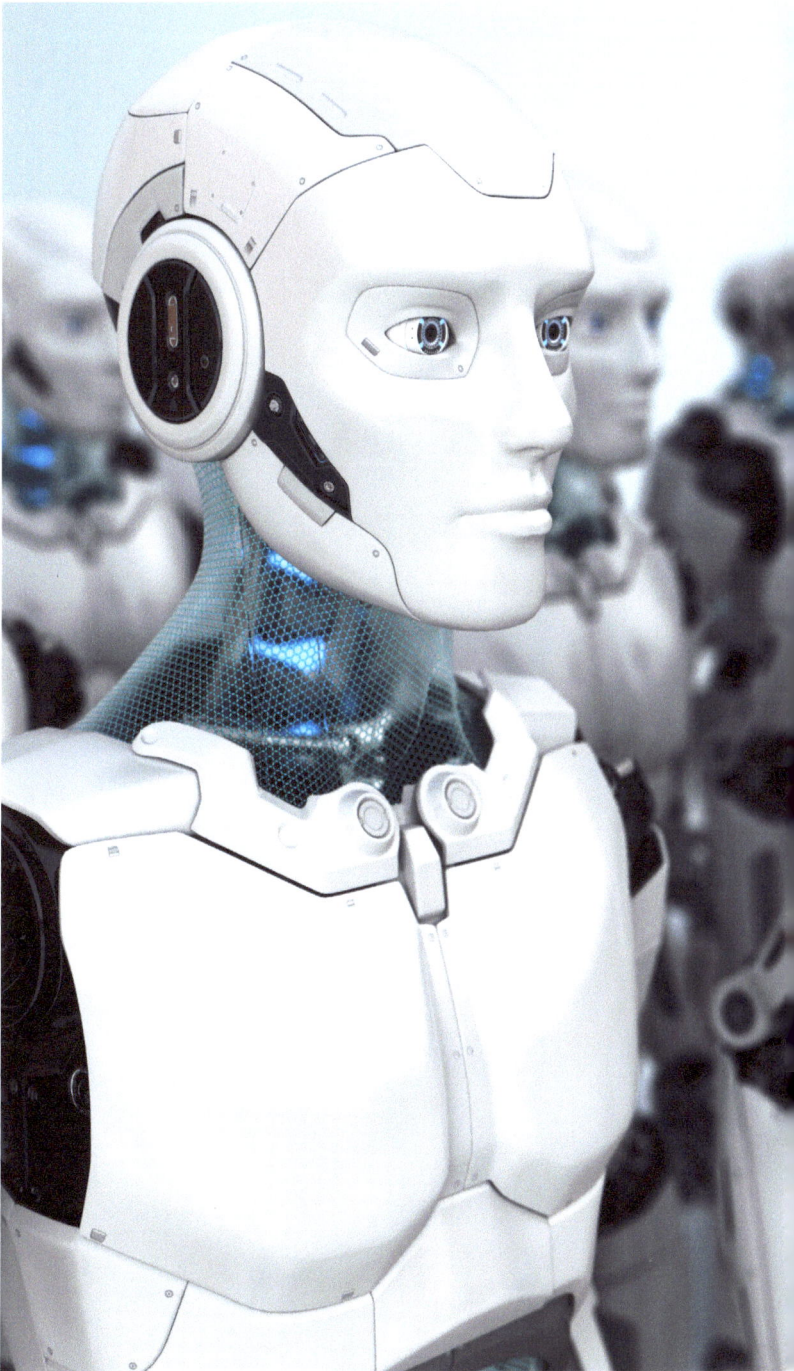

Future Human Role

Future jobs are all about collaboration between smart robots, new technologies, and human.

Human still has a lot of advantages compared to robots and technologies.

In the future, human is going to use the most advanced computer in the world **"Human Brain"** more and more.

Over the past few years, scientists have tried in a variety of ways to get a supercomputer to copy the complexity and to process the power of the human brain.

In a simulation study, scientists needed more than 82,000 processors running on one of the world's fastest supercomputers to simulate just 1 second of a normal human brain's activity.

More recently, a research study found that the human brain could hold 10 times as much information as previously believed.

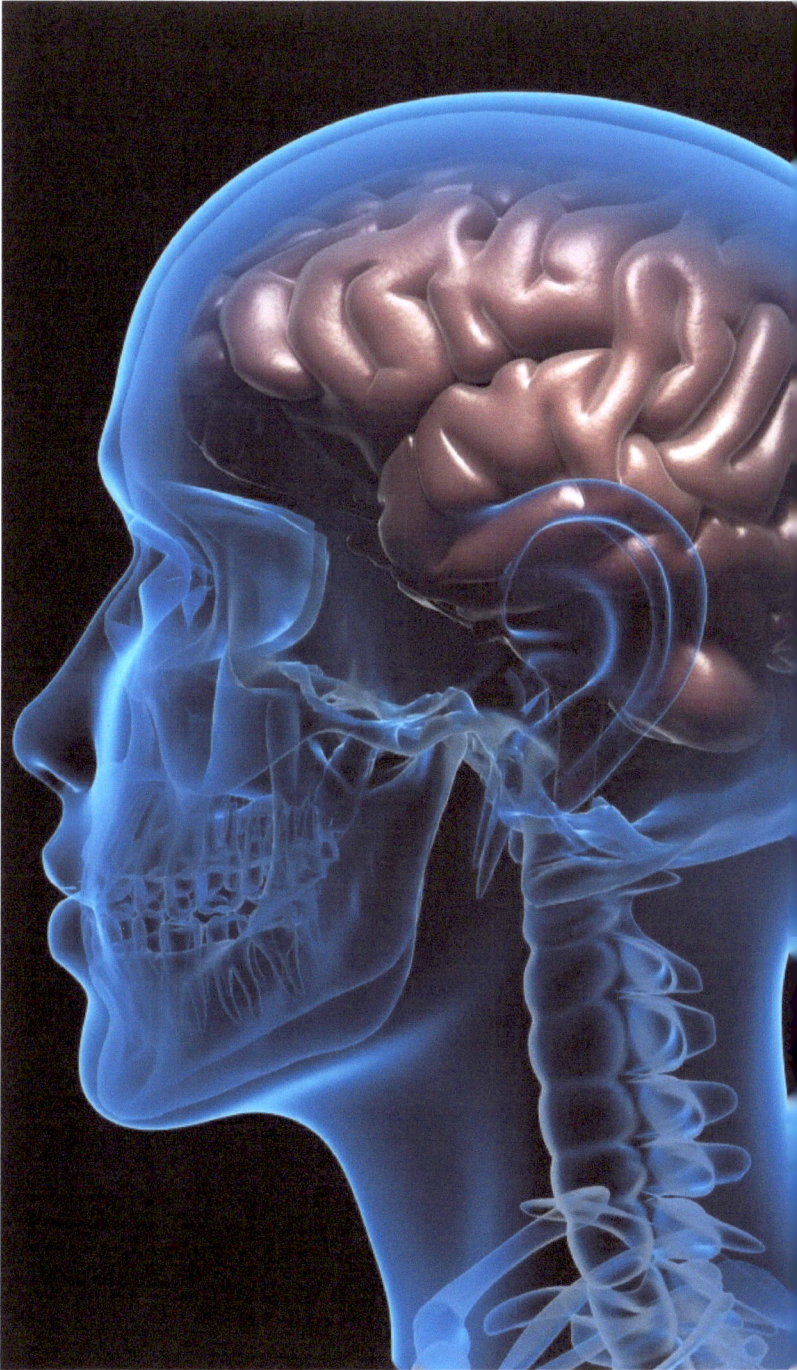

Scientists now believe that the capacity of the human brain is about a petabyte[3].

Brains are also about 100,000 times more energy-efficient than computers. Our brain is:

- The most powerful computer in the world
- The lightest computer
- The most intelligent planner in the world

Creativity, planning, designing, innovating are the areas that human is good at.

Strategic planning, designing products, improving businesses are unique capabilities for human, and machines won't be able to reach the same level anytime soon.

---

[3] 2^{50} bytes; 1024 terabytes, or a million gigabytes.

Last Note

Quality 4.0 references the future of quality and organizational excellence within the context of Industry 4.0.

Quality professionals can play a main role in leading their organizations by using technological tools.

It is also the perfect time for experienced quality professionals to start their new careers as consultants, coaches or educators.

I hope this guideline helps you to get some information about the changes in the future and wish you all the best in your journey!

I would appreciate it if you could leave me a review on Amazon or any other platforms that you have order this book.

To your success!

Sean Shakeri

Future Resources

CLARITY
THE SECRET TO SUCCESS

SEAN SHAKERI

The Step by Step Guide For Your Success

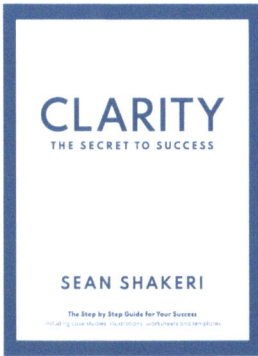

How to organize your life and business
An integrated business and personal
development book

Learn industry 4.0 soft skills
A unique coaching and consulting
program for career improvement

KITINMOTION.COM

References

1.
Sean Shakeri, Clarity – The secret to Success (2020)

2.
Boston Consulting Group, ASQ and German quality of association studies in 2019

3.
World Economic Forum report 2019

4.
Mackinsey Global institute studies 2015–2020

5.
Studies by KIT IN MOTION Inc. 2017–2020

6.
https://en.wikipedia.org/wiki/Fourth_Industrial_Revolution

7.
https://asq.org/quality-resources/quality-4-0

8.
Lucas, Robert E., Jr. (2002). Lectures on Economic Growth. Cambridge: Harvard University Press.

www.ingramcontent.com/pod-product-compliance
Lightning Source LLC
Chambersburg PA
CBHW041118210326
41518CB00032B/194